Separations Division 2017

AA009292

Core Programming Area at the 2017 AIChE Spring Meeting and
13th Global Congress on Process Safety

San Antonio, Texas, USA
26 – 30 March 2017

ISBN: 978-1-5108-4142-0

Printed from e-media with permission by:

Curran Associates, Inc.
57 Morehouse Lane
Red Hook, NY 12571

Some format issues inherent in the e-media version may also appear in this print version.

Printed by Curran Associates, Inc. (2017)

For permission requests, please contact AIChE
at the address below.

AIChE
120 Wall Street, FL 23
New York, NY 10005-4020

Phone: (800) 242-4363
Fax: (203) 775-5177

www.aiche.org

Additional copies of this publication are available from:

Curran Associates, Inc.
57 Morehouse Lane
Red Hook, NY 12571 USA
Phone: 845-758-0400
Fax: 845-758-2633
Email: curran@proceedings.com
Web: www.proceedings.com

TABLE OF CONTENTS

479604 Distillation 101

Abstract:

This paper deals with a short history, and the basics of distillation as a Unit Operation. Distillation theory is converted into practice and how to optimally simulate and choose the appropriate number of stages for a tower.

479808 Distillation & Absorbtion Equipment - Tower Internals

Abstract:

This paper provides a fundamental overview of distillation and absorption tower internals - trays and packings. Fundamentals of tray and packing designs and the basic working of trayed and packed columns will be presented. Photographs and videos from the Fractionation Research experimental unit will be presented.

479809 Operation and Control of Distillation Columns

Abstract:

This paper discusses the fundamentals of the operation and control of trayed and packed distillation columns.

Typical operational issues such as flooding will be discussed.

Control schemes for various distillation column configurations are discussed with their pros and cons.

479804 Energy Savings & Distillation

Abstract:

Distillation is a significant energy consumer in process industries.

Tremendous energy savings have been achieved in recent times thanks to technologies such as heat integration and thermanlly coupled dividing wall distillation columns.

This paper examines the various energy saving schemes for distillation and absorption towers, combinations of other separation technologies with distillation, and the current and future trends in energy efficient distillation.

479812 Acid Gas Absorption Fundamentals

Abstract:

Absorption is a standard unit operation that has been studied by chemical engineering students for decades. One of its primary uses is for the removal of acid gases, like CO_2 and H_2S, from natural gas streams. Historically, aqueous amines and hot carbonate solutions have been used to perform these absorptions. Many different amines, solution strengths, activators and passivators have been tested and employed industrially. Approximately half of the industrial towers are trayed; half, packed. Foaming adversely affects the hydraulic performance in these systems.

Acid gas absorption is receiving appreciable new attention as concerns rise regarding CO_2 in the atmosphere. In addition, the United States has recently experienced significant growth opportunities in shale gas production which require improvements in gas treating technologies. This presentation reviews acid gas absorption and regeneration fundamentals. Several PFD's will be shown to illustrate important concepts. New absorbent solutions will be discussed, including ionic liquids. Flue gas applications will be discussed, including the difficulties associated with column pressure drops and very large column diameters. Rate-based modeling approaches will also be reviewed.

479816 Process Investigations Using Radioisotope Techniques

Abstract:

In all chemical plants and refineries there generally is not enough instrumentation in the right places in our processes. During the design phase, decisions were made between knowing more about the process, which means buying more instruments, and saving capital funds. When a problem appears with the operation of a vessel, there are always gaps in the knowledge of what is happening due to a lack of instruments in the places where knowledge is not routinely required. Process engineers and operations personnel need to know what is happening so they can take corrective action. When knowledge is missing, they often take actions that make matters worse.

For distillation and separation processes Chemical Engineers have historically relied upon plant process measurements such as flow rates, temperatures, and pressures and model or simulation results for data with which to solve troubleshooting or re-design problems. An array of on-line diagnostic services are available that can provide additional data offering real-time information on how pieces of process equipment are actually operating. The most common application is Gamma Scanning. This test is primarily applied, but not necessarily limited to, distillation or separation columns. Gamma scans provide a density profile of the internal process of operating distillation columns and other process vessels. The density profile can be used to diagnose the hydraulic operating conditions of mass transfer devices such as damage to internals, flooding, degree of entrainment or weeping, liquid levels on trays and distributors, liquid distribution through packed beds, etc. The presenter will show case studies where scanning revealed vital process information that helped solve an operating problem, or helped make a revamp/re-design successful. Additional derivatives from gamma scanning of detecting flow distribution patterns will be demonstrated. The ThruVision Scan technique is a specialized horizontal gamma scan used to generate a topographic profile of the internal cross-sectional density of process equipment. This profile is useful for the detailed study of liquid flow distribution through packed columns as well as having applications beyond distillation columns. A hybrid application involving tracers and scanning on fixed bed catalytic reactors will also be discussed.

Another technique involving radioactive materials is the Tracer Test, in which a radioactive tracer is introduced into the process for investigating the process flows. External radiation detectors are placed at strategically chosen locations outside vessels and piping to monitor the tracer flow through the process. Either liquid, gas or a solid tracer can be utilized depending on process conditions and the fluid to be studied. This tracer technique can identify liquid or gas maldistribution in fixed or fluidized bed reactors. The residence time through a tank or reactor can be measured and compared with the calculated residence time. Observed differences can relate directly to the degree of mixing or channeling inside the reactor. These tracer techniques have also been used in identifying leakage of heat exchangers and valves.

NOTES

NOTES

AUTHOR INDEX